YOUR KNOWLEDGE HAS VALUE

- We will publish your bachelor's and master's thesis, essays and papers

- Your own eBook and book - sold worldwide in all relevant shops

- Earn money with each sale

Upload your text at www.GRIN.com
and publish for free

Onkar Sabran

Face detection and tracking using MATLAB

GRIN Verlag

Bibliografische Information der Deutschen Nationalbibliothek:

Die Deutsche Bibliothek verzeichnet diese Publikation in der Deutschen National-
bibliografie; detaillierte bibliografische Daten sind im Internet über http://dnb.d-
nb.de/ abrufbar.

Imprint:

Copyright © 2014 GRIN Verlag GmbH
Druck und Bindung: Books on Demand GmbH, Norderstedt Germany
ISBN: 978-3-656-67766-6

GRIN - Your knowledge has value

Der GRIN Verlag publiziert seit 1998 wissenschaftliche Arbeiten von Studenten, Hochschullehrern und anderen Akademikern als eBook und gedrucktes Buch. Die Verlagswebsite www.grin.com ist die ideale Plattform zur Veröffentlichung von Hausarbeiten, Abschlussarbeiten, wissenschaftlichen Aufsätzen, Dissertationen und Fachbüchern.

Visit us on the internet:

http://www.grin.com/

http://www.facebook.com/grincom

http://www.twitter.com/grin_com

Face Detection and Tracking Using MATLAB

Onkar Nath Sabran
Department of Electronics and Communication Engineering
Mtech in VLSI

Abstract—Face Identification and following has been a vital and dynamic examination field on the grounds that it offers numerous requisitions, particularly in feature observation, biometrics, or feature coding. The objective of this undertaking was to actualize a constant framework on a FPGA board to catch and track a human's face. The face location calculation included shade based skin division and picture separating. The face area was dictated by figuring the centroid of the discovered locale. A product variant of the calculation was autonomously executed and tried on still pictures in MATLAB. Despite the fact that the move from MATLAB to verilog was not as smooth obviously, trial results demonstrated the exactness and viability of the constant framework, much under shifting states of lights, facial postures and skin colors, All estimation of the fittings usage was carried out continuously with negligible computational exertion consequently suitable for force constrained provisions.

I. INTRODUCTION

Face location and following is the methodology of figuring out if or not a face is available in picture. Dissimilar to face distinguishment which recognizes distinctive human confronts, face location just demonstrates whether a face is available in a picture. Moreover, face following decides the precise area of the face. Face identification and following has been a dynamic examination range for quite a while in light of the fact that it is the starting imperative venture in numerous distinctive requisitions, for example, feature observation, face distinguishment, picture upgrade, feature coding, and vitality protection. The materialness of fae identification in vitality preservation is not as evident as in different provision. Not with standing, it is fascinating to figure out how a face identification and following framework permits force and vitality to be spared. Assume one is viewing a TV and dealing with different errands all the while. The face location framework is for checking whether the individual is gazing toward straight forwardly toward the TV.

A. ALGORITHM

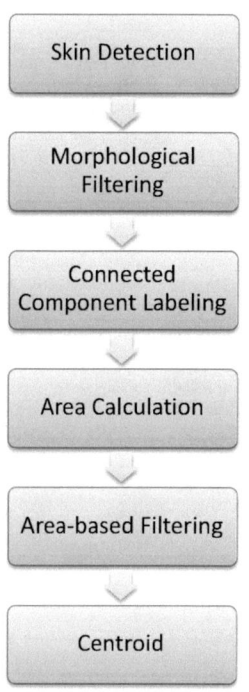

General configuration stages are shown in Figure 1.

The skin location calculation here was determined from the technique describe in [1]. Color division has been turned out to be a compelling strategy to catch face is as because of its low computational prerequisites and simplicity of execution. Contrasted with the emphasized based strategy, the shade based calculation obliged almost no preparation. Initially, the first picture was changed over to an alternate color space, specifically altered YUV. At that point the skin pixels were portioned focused around the proper u reach. Morphological sifting was connected to diminish false positives. At that point each one joined locale of caught pixels in the picture was marked. The territory of each one marked locale was registered and a region based separating was connected. Just districts with extensive region were considered face areas. The centroid of each one face area was like wise processed to demonstrate to its area.

Every present feature casing was caught by the Polaroid and sent to the FPGA's decoder chip by means of a composite feature link. After the feature sign was prepared in distinctive modules in verilog, the last yield passed through the VGA driver to be shown on the VGA screen. The equipment calculation was altered as demonstrated in Figure.

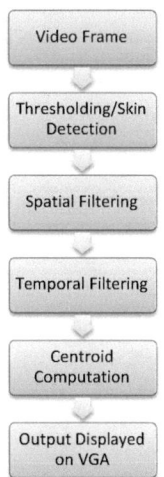

Figure 2 - Hardware Algorithm

A.   Thresholding

Since 10-bit color was utilized within Verilog, modifying the previously stated u reach yields $40<u<296$ in the step, each one info feature casing was changed over to a "parallel picture" indicating the divided crude result.

B.   Spatial filtering

This step was like the disintegration operation utilized as a part of the product calculation. On the other hand, the organizing component utilized here did not have any specific shape. Rather, for each pixel p, its neighboring pixels, p was additionally a skin pixel. Overall p was a non-skin pixel. This permitted most foundation commotion to be evacuated on the grounds that typically clamor scattered arbitrarily through space, as indicated in Figure 9. In Figure 10, in light of the fact that p just had 4 neighboring pixels arranged as skin, p was closed to be a non-skin pixel and, along these lines, changed over to a foundation pixel.

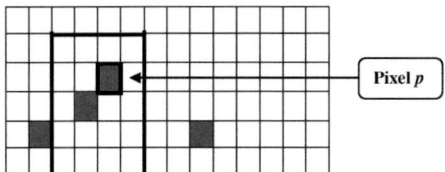

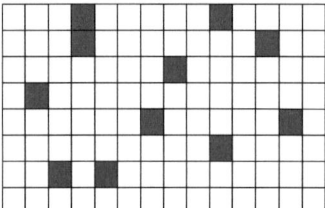

Figure 3 - Example of spatial filtering for a pixel p—before filtering

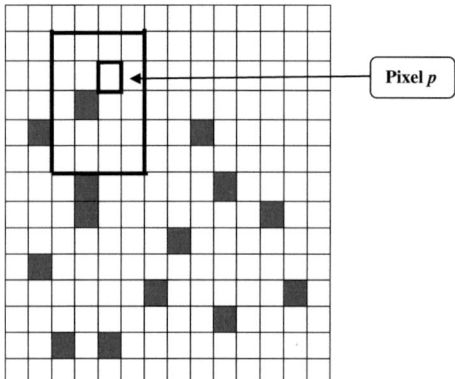

Pixel p

Figure 5 - Example of spatial filtering for a pixel p—after filtering

To look at the neighbors around a pixel, their qualities required to be put away. Thusly, ten movement registers were made to support the estimations of ten back to back lines in each one casing. As seen in Figure 11, each one register was 640-bit long to hold the double estimations of 640 pixels consecutively. Every bit in data_reg1 was upgraded as indicated by the X direction. Case in point, when the X direction was 2, data_reg1[2] was upgraded as indicated by the consequence of thresholding from the past stage. Consequently, data_reg1 was upgraded each clock cycle. After all the bits of data_reg1 were upgraded, its whole esteem was moved to data_reg2. Therefore, different registers (from data_reg2 to data_reg10) were just upgraded when the X direction was 0. Estimations of data_reg2 to data_reg10 were utilized to analyze a pixel's neighborhood.

[	[	[	...	[63	[63	data_reg10
0	1	2	...	8]	9]	
]	]	]				
[	[	[	....	[63	[63	data_reg9
0	1	2	.	8]	9]	
]	]	]				
[	[	[	....	[63	[63	data_reg8
0	1	2	.	8]	9]	
]	]	]				
....						data_reg2

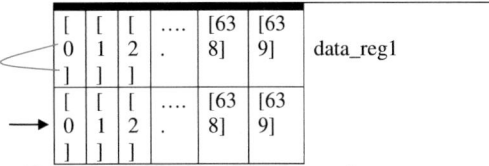

[0]	[1]	[2]	....	[638]	[639]	data_reg1
[0]	[1]	[2]	....	[638]	[639]	

Figure 6 - Ten shift registers for ten consecutive rows

There was an exchange off between the amount of movement registers being utilized (i.e. the span of the neighborhood) and the execution of the spatial channel. A bigger neighborhood obliged more registers to be utilized be that as it may, in the meantime, permitted more commotion to be uprooted.

C. Temporal Filtering

Indeed little changes in lighting could result in gleaming and made the result showed on the VGA screen less steady. Applying transient separating permitted flashing to be lessened essentially. The thought of planning such a channel was obtained from the task "Ongoing Cartoonifier" (see References for more data of this venture). The fleeting channel was focused around the accompanying comparison. avg_out = (3/4) avg_in + (1/4) information information: separated result got from the past phase of a pixel, specifically p, in present casing avg_in: normal estimation of p from past casing avg_out: normal estimation of p in present casing This is give or take equivalent to averaging four back to back casings about whether. To straightforwardness the computational exertion, the comparison above might be re-composed as avg_out = avg_in − (1/4) avg_in + (1/4) information avg_out = avg_in − avg_in >> 2 + information >> 2 The separated consequence of a pixel in this stage was resolved focused around its normal quality (i.e. avg_out). In the event that its normal worth was more excellent than 0.06 (number acquired from examinations), the pixel was considered skin. Generally, the pixel was non-skin. represent the methodology of fleeting sifting for two pixels p1 and p2. In both illustrations, pixel p1 and p2 are really skin pixels. Then again, the results before separating were precarious because of light varieties. The worldly channel smoothed the yield and, consequently, lessened glint fundamentally.

IV. PERFORMANCE

The last come about was a complete framework that was proficient to discover and track appearances of at most two individuals continuously. Despite the fact that it was not ready to track each one face independently when there were three individuals or more, it could in any case identify the vicinity of their countenances.

Analyzes additionally demonstrated that diverse light settings did not altogether modify the last comes about. Besides, the framework could overlook foundation commotion extremely well-generally originated from ligh reflection. At the point when there were questions that had color like skin shade, both spatial and worldly separating helped disintegrate these located locales, in thes way decreasing the amount of false positive.

V. CONCLUSION

In this extend, the objective of executing a fittings framework to recognize and track human faes continuously was attained. A product usage of the calculation was inspected in MATLAB to check its correctness. In spite of the fact that the move from programming to fittings obliged some change to the first calculation, the beginning objective was still finished. The face identification calculation was determined from a skin recognition technique. Face following was accomplished by registering the centroid of each one identified area, despite the fact that it just worked in the vicinity of at most two individuals. Distinctive sorts of channel were connected to abstain from flashing and settle the yield showed on the VGA screen. The framework was demonstrated to work progressively with no slacking and under changing states of facial interpretations, skin tones, and lighting.

REFERENCES

[1] M. Ooi, "Hardware Implementation for Face Detection on Xilinx Virtex-II FPGA Using the Reversible Component Transformation Color Space," in Third IEEE International Workshop on Electronic Design, Test and Applications, Washington, DC, 2006.

[2] S. Paschalakis and M. Bober, "A Low Cost FPGA System for High Speed Face Detection and Tracking," in Proc. IEEE International Conference on Field-Programmable Technology, Tokyo, Japan, 2003.

[3][Online]. Available: http://www.jwp.se/files/skintone.jpg. [Accessed December 2012].

[4][Online].Available:http://people.ece.cornell.edu/land/courses/ece5760/FinalProjects/f2010/kaf42_jay29_teg25 /teg25_jay29_kaf42/index.html. [Accessed October 2012].